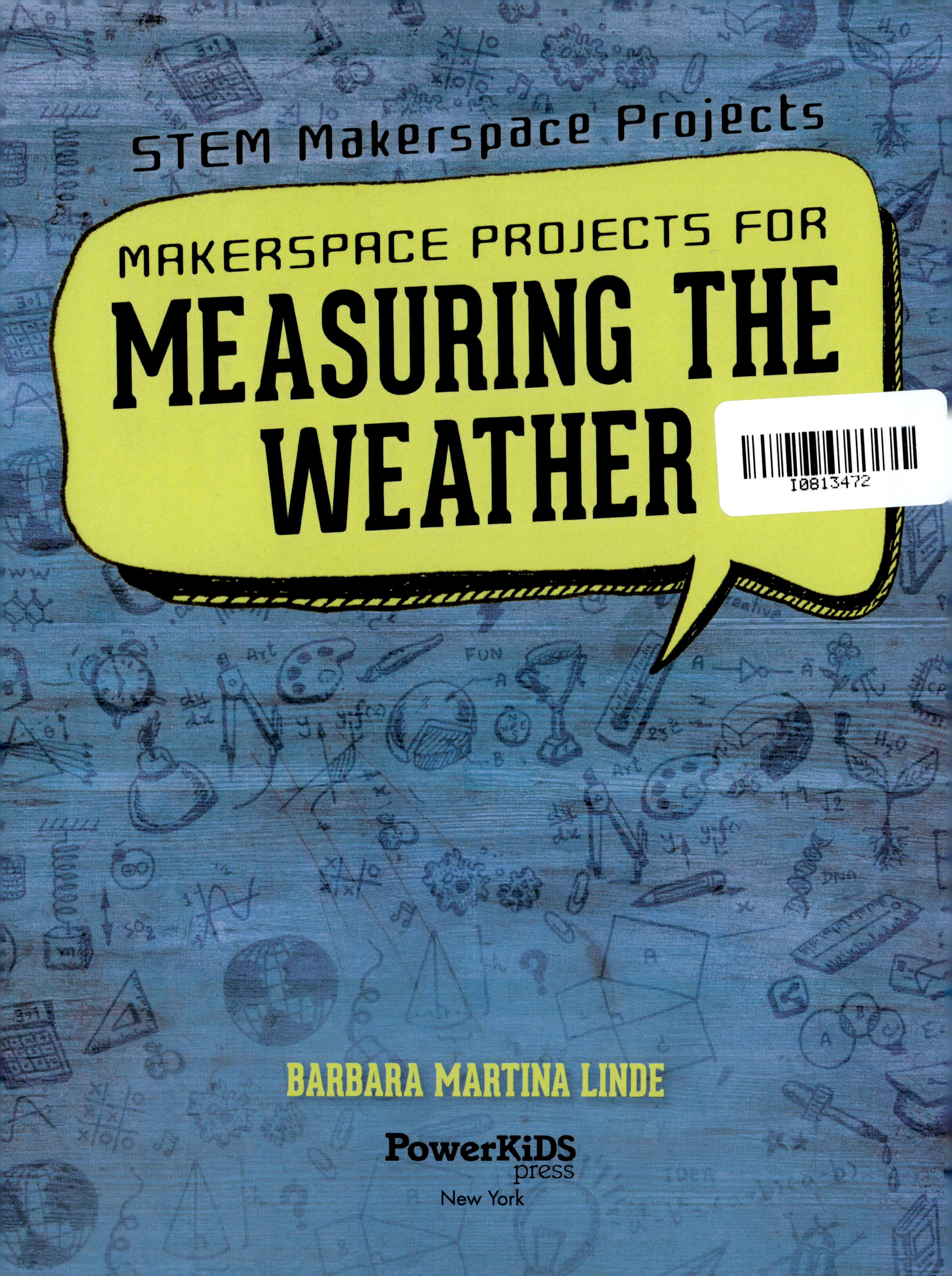
STEM Makerspace Projects
MAKERSPACE PROJECTS FOR
MEASURING THE WEATHER
I0813472
BARBARA MARTINA LINDE
PowerKiDS press
New York

Published in 2021 by The Rosen Publishing Group, Inc.
29 East 21st Street, New York, NY 10010

First Edition

Editor: Danielle Haynes
Illustrator: Benjamin Humeniuk
Book Design: Reann Nye

Photo Credits: Series art (background) ShutterStockStudio/Shutterstock.com; cover Arthorn Saklang/Shutterstock.com; p. 5 Nuei57/Shutterstock.com; p. 7 (snow) Jon Paul Perry arranginglight.com/Moment/Getty Images; p. 7 (rain) ND700/Shutterstock.com; p. 7 (overcast) Quality Stock Arts/Shutterstock.com; p. 7 (sunny) LeManna/Shutterstock.com; p. 9 Merydolla/Shutterstock.com; pp. 11, 15 (barometer) Science & Society Picture Library/SSPL/Getty Images; p. 13 Mette Fairgrieve/Shutterstock.com; p. 15 (Fitzroy) Royal Geographical Society/Royal Geographical Society (with IBG)/Getty Images; p. 16 Daniel J. Rao/Shutterstock.com; p. 17 NoPainNoGain/Shutterstock.com; p. 23 attaphong/Shutterstock.com; p. 24 Mimadeo/Shutterstock.com; p. 25 VRVector/Shutterstock.com; p. 29 S_Photo/Shutterstock.com.

Library of Congress Cataloging-in-Publication Data

Names: Linde, Barbara M., author.
Title: Makerspace projects for measuring the weather / Barbara Martina Linde.
Description: New York : PowerKids Press, 2021. | Series: STEM makerspace projects | Includes index. | Summary: "Will you get a snow day tomorrow? Or are you hoping for a sunny day for a summer picnic? Readers of this volume will learn about weather forecasting and how to make their own weather instruments with three engaging makerspace projects. They'll also learn more about people and events in the history of weather forecasting and the importance of tracking the weather. Budding scientists will enjoy following instructions in the text to create and use their own rain gauge, anemometer, and thermometer. Colorful photographs, diagrams, and graphic organizers will help students visualize the concepts. Fact boxes and sidebars provide additional information to aid reader comprehension"– Provided by publisher.
Identifiers: LCCN 2019032640 | ISBN 9781725311701 (paperback) | ISBN 9781725311725 (library binding) | ISBN 9781725311718 (6 pack) | ISBN 9781725311732 (ebook)
Subjects: LCSH: Weather forecasting–Juvenile literature. | Meteorology–Juvenile literature. | Makerspaces–Juvenile literature.
Classification: LCC QC995.43 .L56 2020 | DDC 551.63–dc23
LC record available at https://lccn.loc.gov/2019032640

Manufactured in the United States of America

CPSIA Compliance Information: Batch #CSPK20. For Further Information contact Rosen Publishing, New York, New York at 1-800-237-9932.

CONTENTS

MEET A MAKERSPACE 4
WEATHER 6
WEATHER AND CLIMATE 8
THE FIRST WEATHER INSTRUMENTS 10
MODERN TOOLS 12
THE FIRST FORECASTER 14
ALL ABOUT RAIN 16
PROJECT 1: RAIN GAUGE 18
ALL ABOUT TEMPERATURE 20
PROJECT 2: THERMOMETER 22
ALL ABOUT WIND 24
PROJECT 3: ANEMOMETER 26
A HOME WEATHER STATION 28
WHATEVER THE WEATHER 30
GLOSSARY 31
INDEX 32
WEBSITES 32

MEET A MAKERSPACE

Welcome to your laboratory! You may be at home or in a classroom. Perhaps you're in a park or a library. No matter where you are, you're in the right place to learn by reading and by doing.

Using this makerspace book, you'll find more than just information about how people measure weather. You'll also find instructions for how to build your very own tools to take measurements. You'll act like inventors of the past and scientists and engineers of the present.

Learning by doing includes your brain and your body. This turns learning into an experience and makes things easier to remember. It also gives you more time to think and ask questions. And it's loads of fun! Are you ready to get started?

What Are Makerspaces?

A makerspace is a community place in which to experiment, build, and create. It could be in a classroom, a garage, or a local library. Any place that has tools, materials, and the freedom to try different ideas can be a makerspace. Makerspaces give people a chance to be curious and creative. You never know what will happen when you spend time in a makerspace!

Look for makerspaces near where you live and go to school.

WEATHER

Weather is the condition of the **atmosphere** over a short period of time. Is it hot and sunny today? Was it cool and cloudy yesterday? Will it snow or rain tomorrow? These are all different kinds of weather.

Weather is made up of six main parts. Temperature tells how hot or cold it is. The weight of the atmosphere is called atmospheric pressure. Wind is moving air. The amount of water vapor in the air is the humidity. Rain, snow, and hail are kinds of **precipitation**. The amount of clouds in the air is called cloudiness. Changes in the air around us cause different kinds of weather. Because this air moves freely all across the planet, weather can change quickly and often.

MAKER MAGIC

The atmosphere is made up of gases such as oxygen and nitrogen. It has six layers and is over 6,000 miles (9,656 km) thick. The atmosphere acts like a blanket around Earth.

Weather can change from hour to hour and day to day.

WEATHER AND CLIMATE

Weather patterns over long periods of time are called climate. The climate of a place can be easy to **predict**, because the seasons repeat every year. Some parts of the year are usually hot. Others are cold and rainy. The weather, though, changes every week, day, or hour. Even today, our best **meteorologists** can't be sure what the weather will be like tomorrow.

Long ago, humans used natural signs to predict the weather. They checked the color of the clouds and smelled the air. They watched animals, which may act differently when a weather change is coming. People read the stars for signs. Some of these methods worked, but others didn't. People always wanted to predict the weather, so they kept trying new things.

Farmers Depend on the Weather

A strong weather event at the wrong time can ruin a farmer's day or year. A cold snap in the early spring can kill young plants. A heavy rain at harvesttime might drown crops. Farmers depend on good weather forecasts to time their activities. For centuries, farmers have used guides called almanacs to help them know the best times to plant and harvest their crops.

An old weather saying is, "Red sky at morning, sailors take warning." Red color meant that the sunlight was reflecting off rain clouds and that the sailors might encounter rain.

THE FIRST WEATHER INSTRUMENTS

Over time, people began studying the weather in more detail. They wanted to keep exact records using numbers. Inventors built instruments to make and keep measurements. In 1441, Korean King Sejong and his scientists invented a new kind of rain **gauge** and a new way to use it. Users operated each gauge exactly the same way. The devices showed measurements the same way, so users knew their work was correct.

Each new invention helped people learn more about the weather. The anemometer measures wind speed. A hygrometer measures humidity. The barometer measures air pressure. In 1709, Daniel Fahrenheit invented a new kind of thermometer to measure temperature. In 1742, Anders Celsius invented a new way to record temperature. Fahrenheit and Celsius measurements both are used today.

MAKER MAGIC

The Greek thinker Aristotle wrote the first known book about weather around 340 BC. He titled it *Meteorologica*. That meant "the study of everything related to the sky." Today, people who study weather are still called meteorologists.

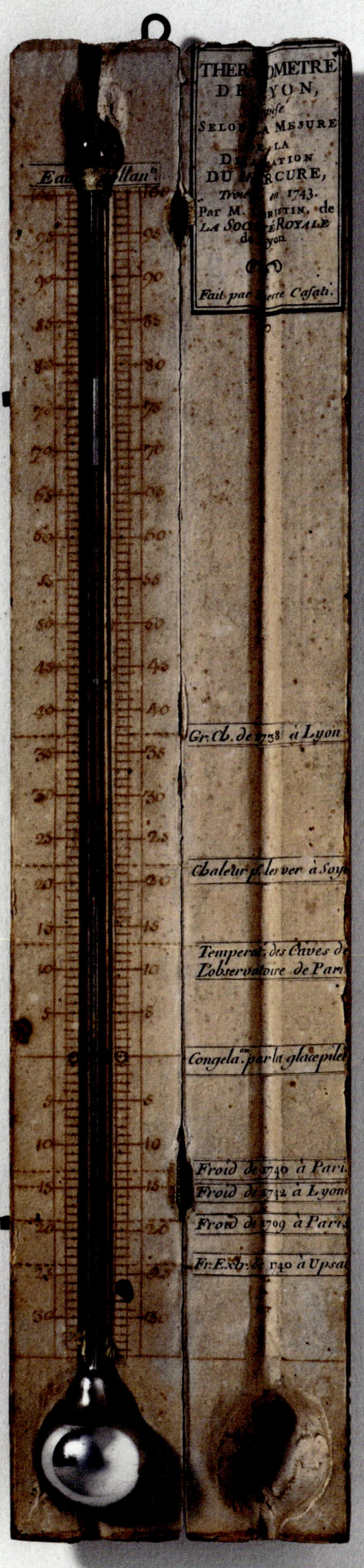

This Celsius thermometer was made by French scientist Pierre Casati around 1790.

MODERN TOOLS

Modern weather-related tools are smaller and easier to move around than early ones. Users can take measurements from more places. Weather balloons carry devices called radiosondes high up into the air. They send the information back to the ground. **Radar** tells us the direction and speed of hurricanes and other storms so forecasters can warn people about dangers. **Satellites** high above Earth look down at weather conditions. Telephones and the Internet help share the information all around the world.

Today's tools use computers, so they're more **accurate**. The computers record weather measurements. Then, computer models combine all the **data**. Computers use the data to tell us what is probably going to happen next. Even so, we can never be totally sure!

MAKER MAGIC

Scientists from different countries began using the Latin language to write about their work in the 1500s. Using the same language helped them better understand each other.

The battery on a radiosonde lasts about two hours.

Radiosonde: Flying Weather Station

A radiosonde is a small box with weather instruments. Scientists send radiosondes up into the air with a balloon, which pops after the radiosonde reaches about 20 miles (32 km) up. A parachute opens to help the box return to Earth without breaking. On its way up and down, the radiosonde collects and sends information. Hundreds of weather observation stations around the world launch radiosondes every day.

THE FIRST FORECASTER

An English man named Robert Fitzroy made the first modern weather predictions in the mid-1800s. He also made up the word "forecast" to refer to such predictions. Before then, more than a thousand sailors died every year because of storms at sea. Fitzroy, a former sailor, wanted to give them warnings so they could stay safe. Weather stations around England used telegraphs to send him information. He used it to warn people about coming storms. Later, he realized he could also predict good weather. Then he started sharing his forecasts with newspapers. Many people used his forecasts.

Fitzroy had been a naval officer before he became a meteorologist. He was also the captain of the *Beagle*, the ship on which naturalist Charles Darwin developed his theory of evolution.

To warn sailors about potentially dangerous weather, Fitzroy had barometers installed at ports throughout the United Kingdom. Some can still be found in special cases at their original locations.

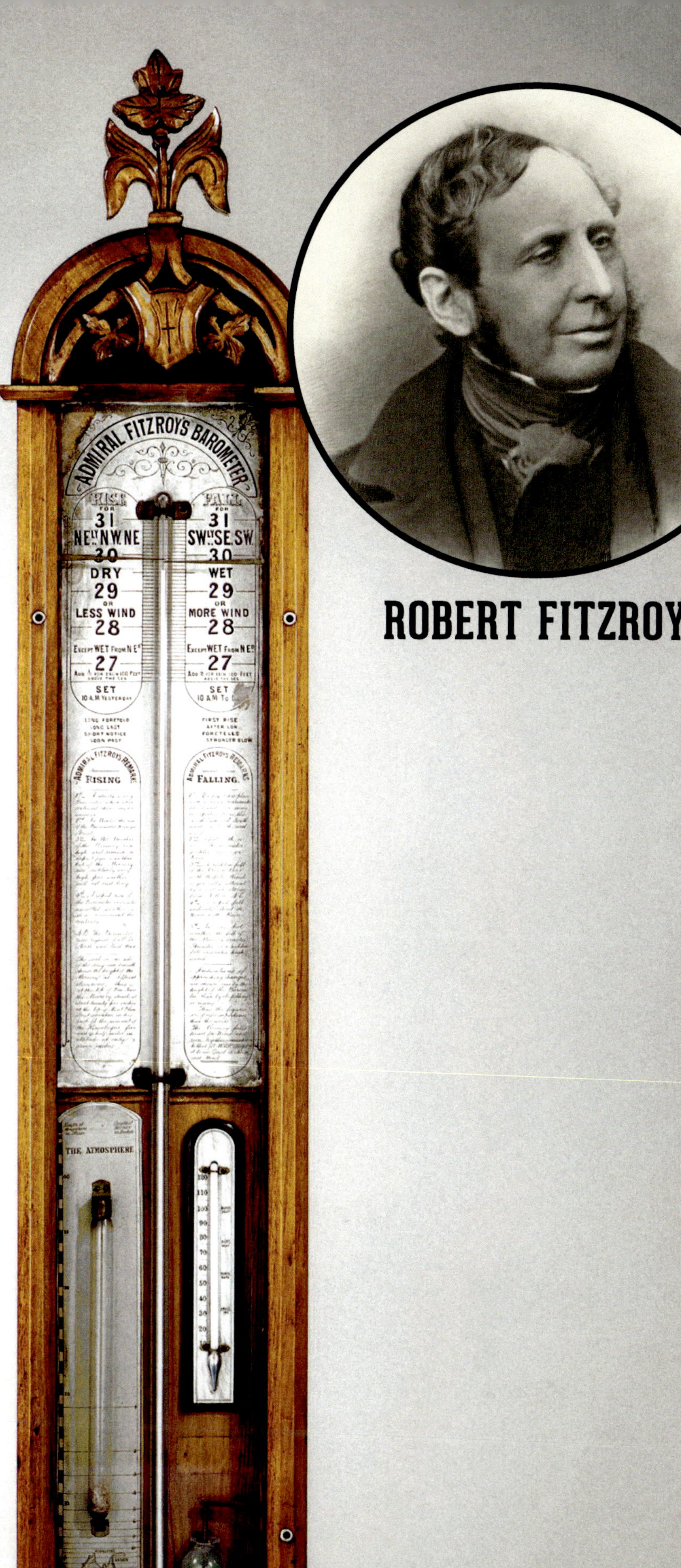

ROBERT FITZROY

ALL ABOUT RAIN

Rain is when water droplets fall out of the clouds. When warm air rises into the atmosphere, it cools. The water vapor in the air begins to form droplets. Inside of clouds, small droplets combine to make big ones. When the big ones get too heavy, they fall down to the ground as rain. Plants, animals, and people all depend on rainwater in order to drink, grow, and be healthy.

MAKER MAGIC

Some deserts receive less than 1 inch (2.5 cm) of rain per year, while the rainiest places get more than 3 feet (91.4 cm) of rain! The rainiest place in the world, Cherrapunji, India, received more than 75 feet (22.9 m) of rain in 1861!

In the water **cycle**, water moves from the air to the land and back to the air again.

Meteorologists measure rainfall for different reasons. Comparing rainfall to the growth of plants lets you know how much water plants need. Comparing this year's rain to last year's lets you know how the climate is changing. Knowing how much rain will probably fall can help you make plans.

PROJECT 1:
RAIN GAUGE

If you want to start measuring the weather and making your own forecasts, where should you start? How about with one of the oldest and most basic of our weather tools—the rain gauge? Rain gauges are simple tools used to measure the amount of rainfall.

WHAT YOU NEED

- 2-liter plastic bottle
- permanent marker
- ruler
- scissors
- small rocks
- tape

WHAT YOU WILL DO

STEP 1:
Draw a straight line around the bottle just under the curved top part.

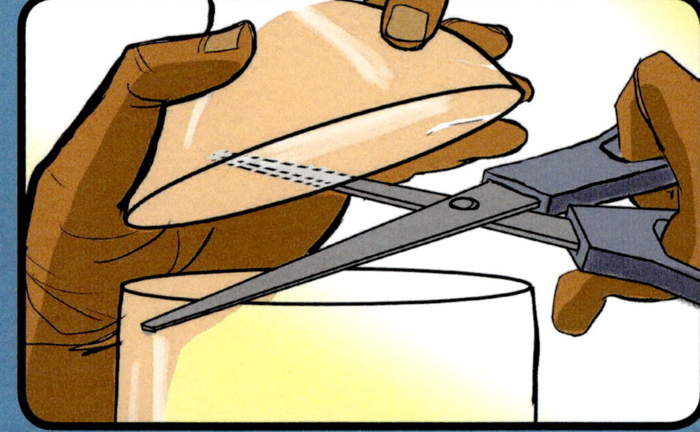

STEP 2:
Use scissors to cut along the line and remove the top of the bottle. Save the top for later.

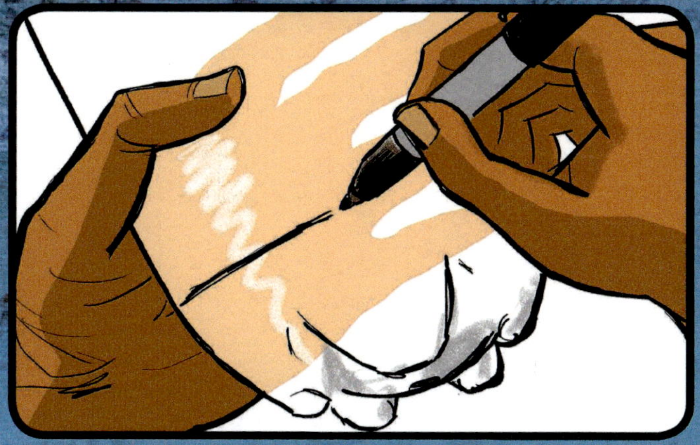

STEP 3:

Draw a straight line around the bottom of the bottle just above where the bottle starts to curve under.

STEP 4:

Starting at the line you drew in step 3, measure up one inch and mark a line. Label it "1 inch." Continue doing this up the side of the bottle until there's no more room (and label "2 inches," "3 inches," etc.).

STEP 5:

Fill the bottom of the bottle with small stones until it reaches the line you drew in step 3. Then pour water into the bottle until it, too, reaches the line. This will help weigh the bottle down and give your rain gauge a level start for measurements.

STEP 6:

Remove the cap. Turn the top part of the bottle upside down. Place it inside the bottom part of the bottle. Line up the top edges and tape the parts together.

STEP 7:

Place the rain gauge in an area that's not covered by trees or a roof. You may need to support the bottle so it doesn't blow away.

STEP 8:

On the next rainy day, check your rain gauge to see how much rain has fallen!

ALL ABOUT TEMPERATURE

When someone asks you how the weather is, you might first tell them whether it's hot or cold. But what if they ask you *how* hot or *how* cold it is? The number on the thermometer you use to answer them might depend on where you live.

In and around the United States, people measure temperature in degrees Fahrenheit. Daniel Fahrenheit was the first person to create **consistent**, accurate thermometers. He also created a temperature scale in which water freezes at 32° and boils at 212°. About 30 years later, Anders Celsius came up with a different scale. On the Celsius scale, water freezes at 0° and boils at 100°. Most countries around the world use the Celsius scale.

MAKER MAGIC

There's also a third scale, used mostly by scientists. It's called the Kelvin scale. It can measure the very low and high temperatures found in outer space. Zero on the Kelvin scale is called absolute zero. This is the lowest temperature possible in the universe. Zero Kelvin is about -273°Celsius and -459°Fahrenheit.

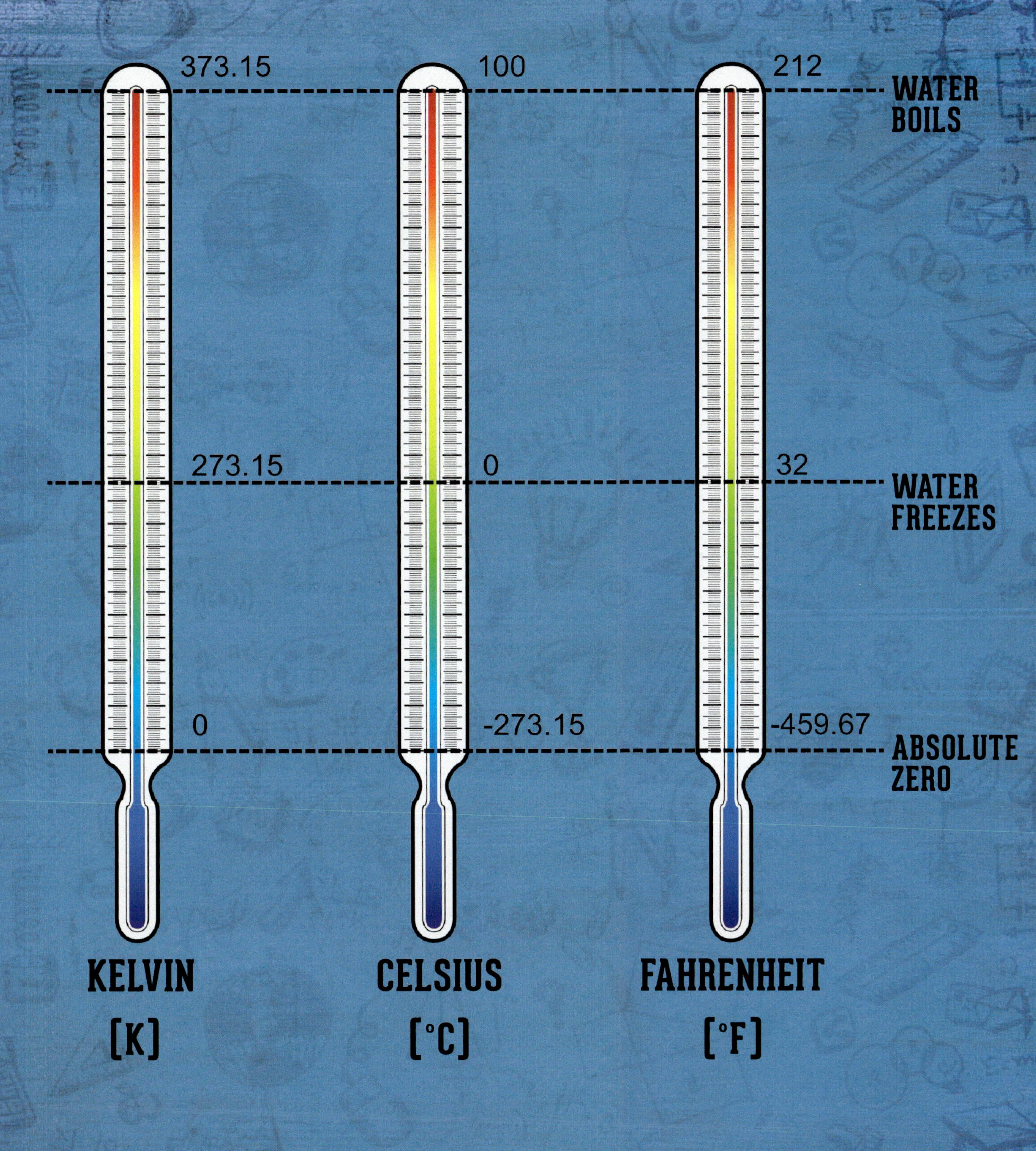

The same temperature is shown using three different scales.

PROJECT 2: THERMOMETER

A thermometer must have a special liquid, such as alcohol or mercury, inside it. The liquid will **expand** or contract when the temperature changes. Here's how you can make a thermometer yourself!

WHAT YOU NEED

- a clear drinking straw
- a small plastic or glass bottle, washed and dried
- a ruler
- a marker
- modeling clay
- water
- rubbing alcohol
- a few drops of food coloring
- an eyedropper
- one small bowl of hot water
- one small bowl of ice water
- one store-bought thermometer

WHAT YOU WILL DO

STEP 1:

Use the marker and ruler to draw lines on the drinking straw every ¼ inch.

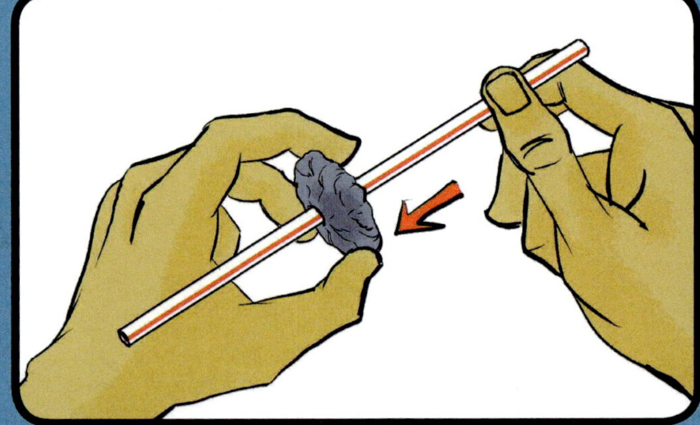

STEP 2:

Use the modeling clay to form a flat, round shape a little bigger than the mouth of the bottle. Use the straw to poke a hole in the clay.

STEP 3:

Fill the bottle halfway with rubbing alcohol and add a few drops of food coloring. Mix well.

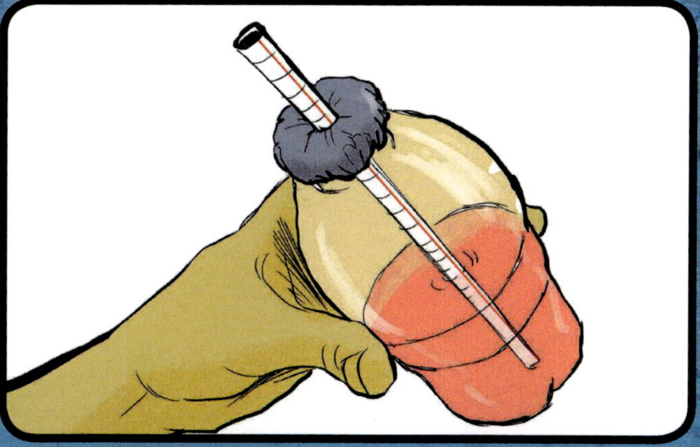

STEP 4:

Cover the opening of the bottle with the clay. Put the straw through the hole you poked into the clay in step 2. The straw should sit low in the liquid inside the bottle, but it shouldn't touch the bottom of the bottle. Seal the clay around the opening of the bottle and around the straw so it's airtight.

STEP 5:

Using the eyedropper, add more alcohol into the straw until the level is above the modeling clay. You'll know the clay seal at the top of the bottle isn't airtight if the level of alcohol in the straw doesn't rise and stay up.

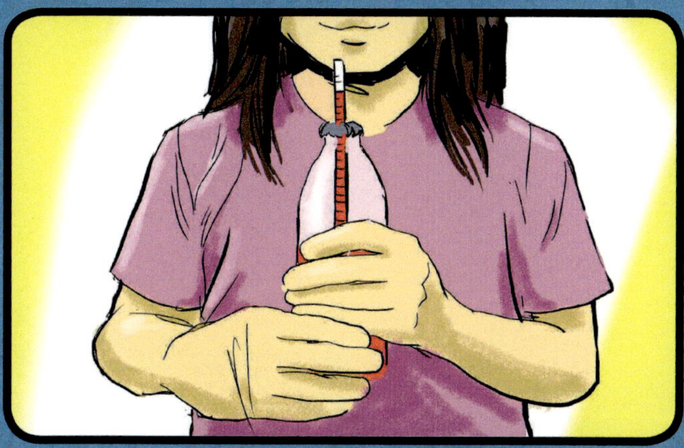

STEP 6:

Hold the bottle in your hands for a few minutes. Your body temperature will heat the alcohol. When it gets warmer and expands, the level in the straw will increase.

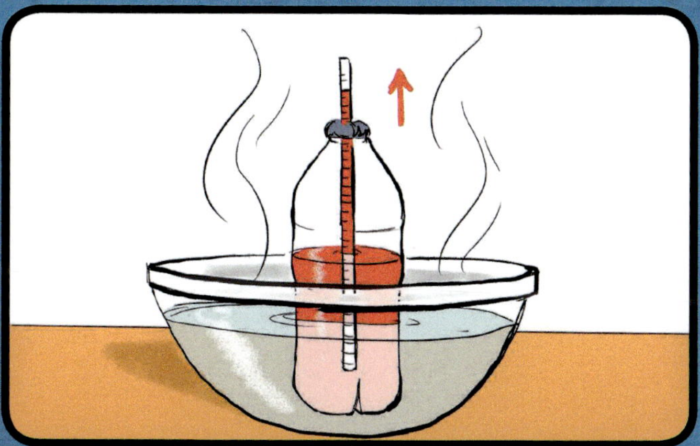

STEP 7:

Hold the bottle in a bowl of hot water. It will expand even more!

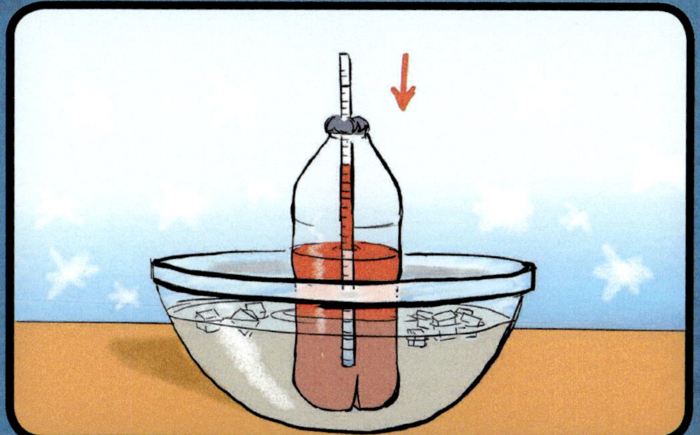

STEP 8:

Hold the bottle in a bowl of ice water. Watch the level drop!

ALL ABOUT WIND

Have you felt a chilly draft or a pleasant breeze? How about a sudden gust? All of these are types of wind!

Wind is air moving from areas of high pressure to areas of low pressure. As the sun heats up parts of Earth, the air there rises. When it does, it leaves a low-pressure area behind. Other air comes in to make the pressure equal, creating wind.

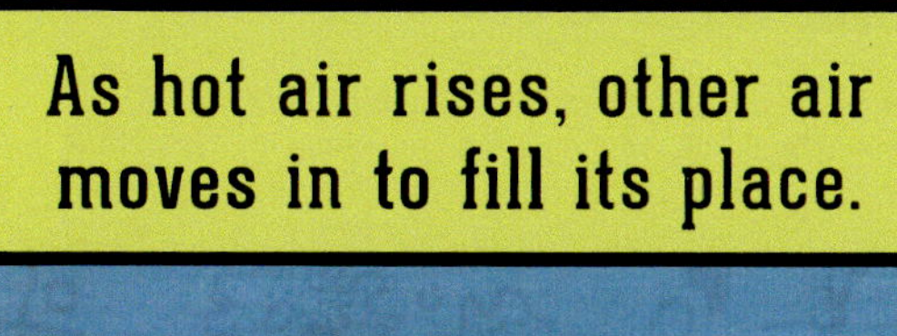

HOT AIR

COOL AIR

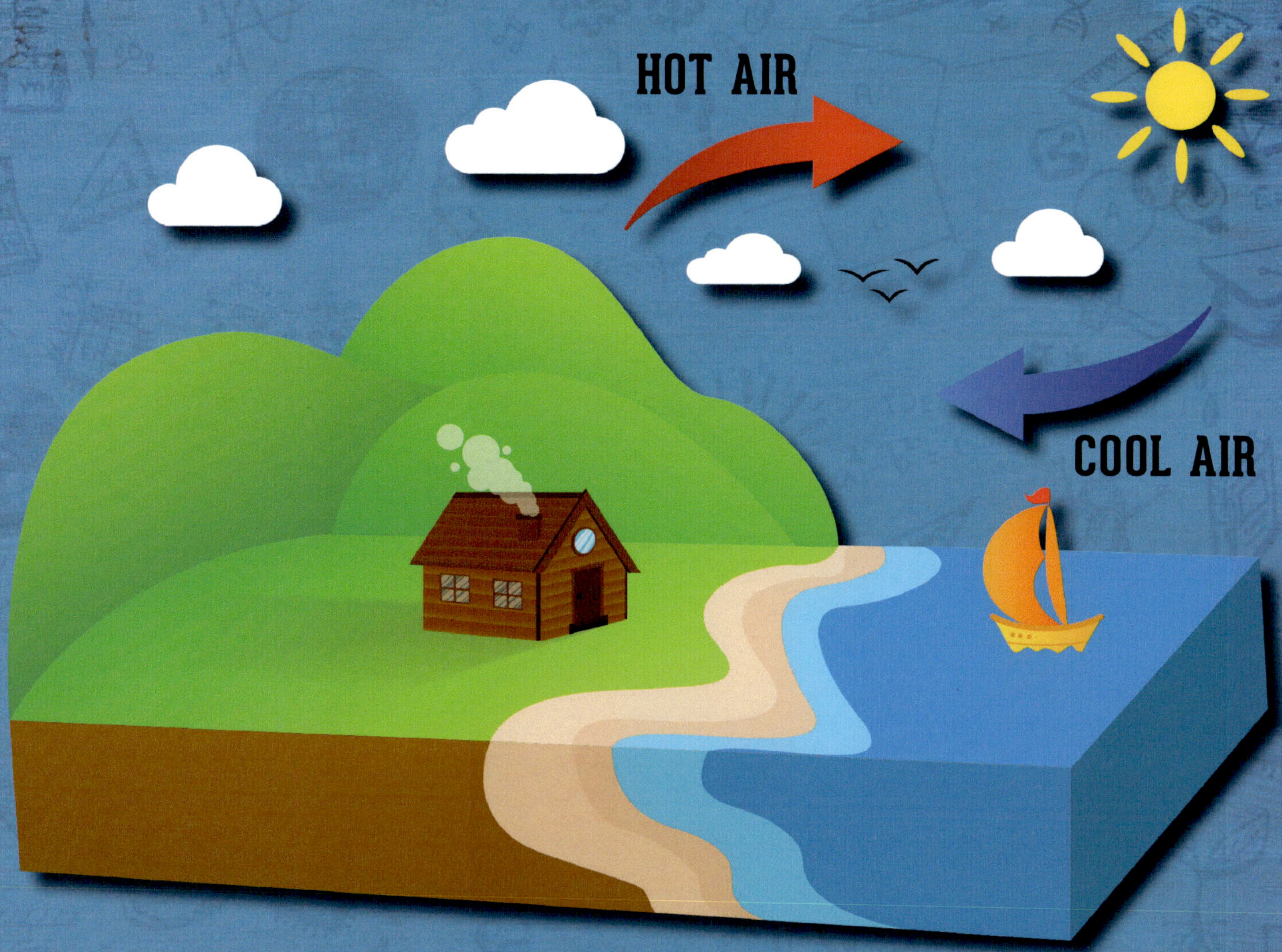

We can measure the wind to find its speed and its direction. The direction of the wind tells us which way storms might move. The speed tells us how strong the winds are. Strong winds can be dangerous for ships, crops, buildings, and people. Spots that frequently have strong winds would be good for wind **turbines**. These harness wind energy.

PROJECT 3:

ANEMOMETER

You may not be able to see the wind, but you can smell it, you can feel it, and you can definitely measure it! Wind speed is measured using a tool called an anemometer. These usually have cups to catch the wind. The wind pushes them around in a circle. Counting how many times they go around will tell you how fast the wind is blowing.

WHAT YOU NEED

- five paper cups
- two plastic straws
- one pushpin
- one pencil with eraser
- hole puncher
- tape
- markers or colored pencils
- watch, stopwatch, or other timer

WHAT YOU WILL DO

STEP 1:

Take one paper cup for your base. Punch four holes just below the lip of the cup. There should be equal space between each hole so each is exactly opposite one of the other holes.

STEP 2:

Stick one straw through two of the holes that are opposite each other. Repeat with the second pair of holes so you make a cross with four segments sticking out.

STEP 3:

Take another cup and punch a hole about ½ inch below the rim. Do this with all the cups. Color one cup differently from the other three.

STEP 4:

Push one of the straws sticking out from the base into a hole in one of the cups. Bend about ½ inch of the straw over and tape it inside the cup. Do this with the rest of the cups and straws. Make sure the opening of each cup faces the bottom of the cup next to it so that all cups point in the same direction as they spin around the base cup.

STEP 5:

Turn the base cup over so the bottom is up and use the pencil to poke a hole in the center of the bottom.

STEP 6:

Now turn the pencil over and insert it eraser-first through the hole.

STEP 7:

Turn the whole thing over again and stick the pushpin through the intersection of the two straws and into the pencil eraser, but make sure it's not too tight!

STEP 8:

Take your anemometer to a windy spot and hold it up. Set a timer for 30 seconds. Count how many times your colored cup revolves during that time. Repeat in other places to find out where the wind is strongest.

A HOME WEATHER STATION

Now you're ready to make your own weather station. Think about what kind of location each device needs.

Your rain gauge needs to be out in the open. It won't catch any rain if it's covered by a roof or a tree. Your thermometer should sit a few feet off the ground. Dirt and concrete on the ground absorb heat, so a thermometer there will give a high reading. Your anemometer should be somewhere in the open. If it's blocked by a wall, it won't catch any wind. But if you put it somewhere high, remember that the wind might blow faster there than on the ground.

Once you've found the perfect spot, set it up. Now you're ready to start recording data yourself!

Use your weather station many times.
Keep track of your data.

WHATEVER THE WEATHER

Well done! You've now built three of your own tools for measuring the weather. It took humans thousands of years to invent these devices. Now that you know how to do it, though, you can create them with simple and inexpensive materials.

The real challenge is to figure out what's next! What else do you want to know? Is there anything else you want to measure? Maybe you'd like to try making a barometer to measure air pressure. Do you want to measure humidity? Make a hygrometer. You could also try taking measurements at your home and school and seeing how they're different. Or compare this year and next year. It's up to you! When it comes to measuring Earth's weather, the sky's the limit!

GLOSSARY

accurate: Free of mistakes.

atmosphere: The mixture of gases that surround a planet.

consistent: Always happening the same way.

cycle: A repeating period of time.

data: Facts and figures.

expand: To become bigger.

gauge: A tool that measures the amount of something.

meteorologist: Someone who studies weather, climate, and the atmosphere.

precipitation: Water that falls to the ground as hail, mist, rain, sleet, or snow.

predict: To guess what will happen in the future based on facts or knowledge.

radar: A machine that uses radio waves to locate and identify objects; a way of using radio waves to find distant objects.

satellite: A spacecraft placed in orbit around Earth, a moon, or a planet to collect information or for communication.

turbine: An engine with blades that are caused to spin by pressure from water, steam, or air.

INDEX

A
anemometer, 10, 26, 27, 28
Aristotle, 10

B
barometer, 14, 30

C
Celsius, Anders, 10
climate, 8, 17
clouds, 6, 8, 9, 16

F
Fahrenheit, Daniel, 10, 20
Fitzroy, Robert, 14, 15

H
humidity, 6, 10, 30
hygrometer, 10, 30

M
meteorologist, 8, 10, 17

R
radiosondes, 12, 13
rain, 6, 8, 9, 10, 16, 17, 18, 19, 28
rain gauge, 10, 18, 19, 28

S
Sejong, King, 10

T
temperature, 6, 10, 20, 21, 22, 23
thermometer, 10, 11, 20, 22, 28

W
water vapor, 6, 16
wind, 6, 10, 24, 25, 26, 27, 28

WEBSITES

Due to the changing nature of Internet links, PowerKids Press has developed an online list of websites related to the subject of this book. This site is updated regularly. Please use this link to access the list: www.powerkidslinks.com/stemmake/weather